HI-TECH JOBS WITHOUT COLLEGE

BE A DATA ANALYST

by Tammy Gagne

BrightPoint Press
San Diego, CA

an imprint of ReferencePoint Press, Inc.
Printed in the United States

For more information, contact:
BrightPoint Press
PO Box 27779
San Diego, CA 92198
www.BrightPointPress.com

LIBRARY OF CONGRESS CATALOGING-IN-PUBLICATION DATA

Name: Gagne, Tammy, author.
Title: Be a data analyst / by Tammy Gagne.
Description: San Diego, CA: ReferencePoint Press, 2026 | Series: Hi-tech jobs without college | Includes bibliographical references and index. | Audience: Grades 7–9
Identifiers: ISBN 9781678212629 (hardcover) | ISBN 9781678212636 (eBook)
The complete Library of Congress record is available at www.loc.gov.

CONTENTS

AT A GLANCE 4

INTRODUCTION 6
MAKING THE NUMBERS WORK

CHAPTER ONE 12
EXPLORING DATA ANALYTICS

CHAPTER TWO 22
TRAINING FOR DATA ANALYTICS

CHAPTER THREE 34
WORKING AS A DATA ANALYST

CHAPTER FOUR 46
LOOKING AHEAD

Glossary 58
Source Notes 59
For Further Research 60
Index 62
Image Credits 63
About the Author 64

AT A GLANCE

- Data analysts collect, review, and interpret data. Then they use this data to solve problems.
- Data analysts are in high demand in many industries. They may work in banking, education, or entertainment.
- Many companies no longer require college degrees for data analysts.
- Data analysts need strong math skills. Analysts must also know one or more programming languages.
- Certificates can be especially helpful for people seeking an analyst position without a college degree. A certificate may also help them get a promotion or earn more money.

- Some data analysts always work in an office. Others work fully from home.
- A data analyst's main tool is a computer. Analysts use software to manage large amounts of data.
- Demand for data analysts is on the rise. This growth is driven by advances in technology.
- Experts do not believe artificial intelligence (AI) will take away data analytics jobs. But as AI does more, human data analysts will need to learn to work with AI-driven systems.

MAKING THE NUMBERS WORK

Avery Smith helps businesses solve problems. He is a data analyst. Analysts gather data. Then they organize it. Data is facts and figures. It includes numbers, text, and images. Analysts explain what the data means. This helps them solve problems.

Smith studies a company's data. He may review sales numbers. He might check to see if ads have increased sales. Sometimes

Data analysts gather data and study trends. This helps businesses make better decisions.

CEMBER
3 4 5 6 7
24
31
75%
100%

Smith looks at how long it takes to ship products. He searches for ways to save money, time, and energy.

Smith has created programs that improve safety. He made a program for

Analysts can help companies provide customers with more accurate shipping times, which can lead to better sales.

Vaporsens. This company builds machines that sense dangerous chemicals. Smith's model is used in airports. It helps detect threats such as explosives.

Smith also helps people become data analysts. He teaches them the skills they need to find one of these jobs. He hosts boot camps. These are intense learning programs. He also hosts a podcast.

Smith teaches the easiest skills first. For example, Smith suggests learning Microsoft Excel. This is a spreadsheet program. Knowing how to use it for data analysis helps people advance quickly. "It just doesn't take a long time, and it's a very valuable tool," he says.[1] Smith believes it takes about 90 days for people to learn the skills needed to become a data analyst.

Data analysts can play important roles in helping companies be more efficient.

THE FIELD OF DATA ANALYTICS

Data analysts are in high demand. They are needed in many industries. These include banking, education, and entertainment. These industries must manage a lot of data.

Data analysts study data. Then they use data to help companies solve problems. This can save money. Or it might lead to a new product customers want to buy. Data is everywhere. Data analysts help others understand it.

Data analysts in the banking industry help banks assess risk and spot trends.

CHAPTER ONE

EXPLORING DATA ANALYTICS

Industries collect data every day. One example is retail. Banks and stores gather data about what customers buy. This includes which goods or services they buy. Where they buy items is important, too. Companies want to know how much customers are spending. This data helps businesses. It can show them which products people want. It can also help them decide how much to charge for items.

In 2025, more than 85 percent of Americans shopped online.

Gathering data is important. But data alone does not give information. Companies need to know how to read the data. This is where data analysts come in.

Data analysts collect and review data. They also interpret it. Then they use this

Data analysts organize data to study trends, see patterns, and look for areas where businesses can cut costs.

data to solve problems. They look for hidden patterns in numbers. Jason Eborn works in data analytics. He explains, “Data analysts are the modern-day detectives of the business world.”[2]

When online sales drop, the company wants to know why. Analysts study the website. They look at recent changes to its layout. They see how long it takes pages to load. Analysts also see which parts of the site are getting clicks. Analysts work to find answers.

Analysts can help with other problems, too. A business may have too few repeat customers. Or it may have too many product returns. Analysts may notice ways to sell more goods. They might have ideas about how to bring down costs. Or they see how employees could save time. The data that analysts gather can help businesses. It can help leaders make better decisions. This can help companies make more money.

THE PROCESS

Data analytics has many steps. Analysts often begin by setting up a data system. This might include databases. Analysts also use other software programs. These tools help analysts collect data. They can help analysts study and report data, too.

Computer code is a list of instructions. It tells computers to do certain tasks. Code is written in different programming languages. All apps and programs rely on coding. Knowing how to code can help

Data analysts use a Python package called Pandas for data organization and analysis.

data analysts. This skill makes it easier to find errors or other problems in the data.

Analysts collect raw data. Then they clean it. This is called data mining. They fix errors. For example, analysts look for typos. The data must also be consistent. A state cannot be listed by its full name in one entry and abbreviated in another. And information should not be repeated.

Next, analysts study the data. They may write reports to explain it. They also give ideas about how it can be used. Presenting data in a way that is easy to understand is key. This may mean creating charts or graphs.

IMPORTANT SKILLS

Data analysts need strong math skills. Knowing statistics is helpful as well. These skills help with gathering and organizing data. They help with measuring data, too. Analysts must also know one or more programming languages. Python, R, and SQL are commonly used by analysts.

Analysts must also know software. Microsoft Excel and Google Sheets are used to make spreadsheets. Tableau and

Qlik are used for **data visualization**. They help analysts create graphics.

Good communication skills are also important. Stephen Baraka is a data analyst. He states:

> *Presenting data analysis is an important part of [the] job. Besides compiling the findings in a clear manner, data analysts also explain*

Not Just Numbers

Data analysts work with numbers. But they might also work with text. For example, users may add comments about products on social media. Users may state what they like about an app. Or they might say what they don't like. Data analysts gather this information. They use the comments to improve the app.

Working on teams allows analysts to share ideas and come up with many possible solutions for clients.

both verbally and in writing why the data is important and what the company can do to respond to the findings.[3]

Data analysts should be good problem solvers. Critical thinking skills are important. Analysts should be good at making decisions. They need to be able to explain why they made the decisions, too.

Data analysts must work well on a team. They will work with programmers and engineers. They will also work with business leaders. An analyst should be open to trying ideas from team members. Attention to detail is important, too.

CHAPTER TWO

TRAINING FOR DATA ANALYTICS

Data analysts once needed a college degree. Some positions still do. But many companies no longer require college. They want people to have the necessary skills. Some people developed these skills in high school. Others have learned in different ways. These include boot camps or self-study courses. Boot camps offer more structure. Self-study courses allow people to learn at their own pace.

Data analytics classes are available in person at universities and community colleges or online through a variety of learning platforms.

Many people learn about data analytics through classes. Some colleges let anyone take classes. Students do not need to be enrolled full time. They can take classes that relate to their career goals. Courses covering statistics and Excel are helpful. Computer programming is needed, too. Classes may be in person or online.

Coursera offers classes from companies such as Google and IBM on basic data analytics concepts and Python.

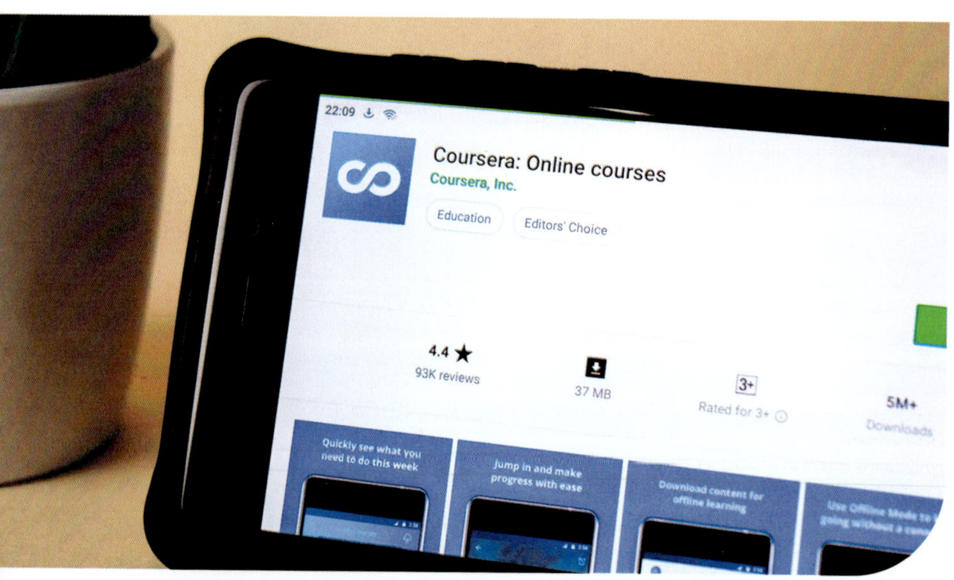

Online platforms are another way to study data analytics. These include Coursera and Udemy. Some classes on these platforms are taught by professors at well-known universities. They are usually cheaper than college courses.

Lauren Rosenthal is a data analyst. She saw a TikTok video about SQL. The video inspired her to learn about data analytics. Rosenthal signed up for an SQL class through Udemy. She says:

> *I took it, enjoyed the content, and wanted to keep learning more. In the following weeks, I went back to Udemy and bought another course, and another, and another. . . I bought courses in Excel and Tableau and SQL and R and Python.*[4]

Boot camp programs in data analytics focus on problem-solving and learning how to use tools to create data visuals.

Students may also take workshops. These classes may be offered at community centers or other local organizations. Some workshops last a few weeks. Others are only one day. Workshops teach different skills. One may teach data cleaning. Another might teach Excel skills. People can build new skills with each workshop they take.

Boot camps are similar to workshops. Both typically offer hands-on experience.

Boot camps may also provide **mentorship**. They may help with job placement. Boot camps tend to be more intense. Classes are held every weekday. Boot camps run from several weeks to months. The classes focus on building skills. Students may learn to code. Or they may study digital marketing. Data analysis is a common course as well.

A SELF-TAUGHT APPROACH

Some people struggle with classroom learning. They prefer learning at their own pace. These people can study data analytics on their own. They may read books about programming languages and software. This can help build a knowledge base. But it is essential to use current sources. Technology changes quickly.

Online videos can be good ways to learn necessary skills such as the programming languages SQL and Python.

People can also learn from instructional videos. These are found on online platforms such as YouTube. Future analysts can find self-paced tutorials, too. These interactive lessons allow users to practice what they learn. Examples of tutorials include W3Schools and The Python Tutorial. Online videos may also teach problem-solving methods. This can help new analysts learn different ways to solve common data analytics issues.

Learning outside the classroom can be more flexible. This means people may continue working while they learn. Students can read or take tutorials on their own time.

Finding New Solutions

Some people leave other jobs to become data analysts. Mike McCulloch works for CareerFoundry. He is the director of career outcomes. He thinks newcomers have some advantages. He says, "They don't yet know what's possible, so they ask different and unexpected questions. Not only does this keep seniors on their toes; it also helps the business to find new solutions to old problems."

Quoted in Emily Stevens, "How to Become a Data Analyst with No Experience or Degree," CareerFoundry, *February 22, 2023. https://careerfoundry.com.*

These options are also affordable. Books may be available through local libraries. Online videos and tutorials can be found at little or no cost.

GETTING CERTIFIED

Professional certificates are another option. Earning a certificate shows that a person has built data analytics skills. Certificates can be earned online. Community education programs may offer them as well. The Google Data Analytics Professional Certificate is an option for beginners. It shows that the person has learned basic SQL. It also shows they know how to use Google Sheets. This is a web-based platform. It allows users to create and use spreadsheets online.

Another option is the Microsoft Certified Power BI Data Analyst Associate certification. It is usually earned by someone who has more experience. It focuses on **data modeling**. Students also learn how to create a dashboard in Microsoft data systems. This is where key information is displayed.

Earning an online certificate in data analytics can be an effective way to stay up to date on the latest advances in the field.

Earning data analysis certificates is one way applicants can demonstrate their interest in this career.

Certificates can be helpful for people who don't have college degrees. Certificates help new analysts learn basic skills. Experienced data analysts earn them, too. A certificate may help them get a promotion. Or it may lead to a raise in pay.

Certifications have their own requirements. For example, users may need experience in analytics. They might also need to complete courses and an exam. Users who pass this test earn the certificate. Becoming certified in data analytics can take time. Completing the Google Data Analytics Professional Certificate usually takes about 6 months. It requires about 10 study hours per week.

There are eight courses required for this certificate program. Tamyris Gimenez took those courses. Then she passed the exam. The entire process took about 8 weeks. She shares, "I had very little knowledge of this field when I first started, and although I do not feel 'job-ready' quite yet, I know a lot more than I expected to in such a short time."[5]

People new to data analysis must learn many skills. Those in the field need to keep learning as well. A lot of training also happens on the job.

CHAPTER THREE

WORKING AS A DATA ANALYST

Agatha Kang is a data analyst. She works for Amazon. Kang uses her skills to solve business problems. Each morning, she checks her email. Messages may be about problems with her current data systems. One morning, she discovered a database had failed overnight. This was caused by a timeout error. These errors may happen when a program takes too

A big company such as Amazon relies on thousands of data analysts to study trends, track customer feedback, and complete other analytical tasks.

amazon

Regular meetings provide opportunities for analysts to gain insights from team members.

long to run. She solved the problem by rerunning the program. This time it worked.

Kang goes to many meetings each week. Some help her learn about specific problems that need to be solved. She can create the best data system for the problem. Other meetings provide updates. She meets with her coworkers and managers. She tells them how her current projects are going.

For most of the day, Kang works on projects. Some are big projects. They are made up of many smaller tasks. She may spend a year or longer on this type of work.

Problem-solving is a big part of her job. Sometimes this is Kang's favorite part of her job. Other times she doesn't like it. She says, "There have even been times where I'm dreaming about data . . . thinking about how I can solve a problem."[6] At the end of the day, she responds to new emails. This helps her start the next day fresh.

TAKING THE WORK HOME

Kang's job at Amazon is a hybrid position. This means she goes to the Amazon office 3 days each week. She works at home the other 2 days. Data analytics can be mentally tiring. Some days, Kang starts in the office. Then she finishes her work at home.

Data analysts may have the flexibility of working remotely at least 1 day per week.

THE BEST PLACES TO FIND WORK IN DATA ANALYTICS

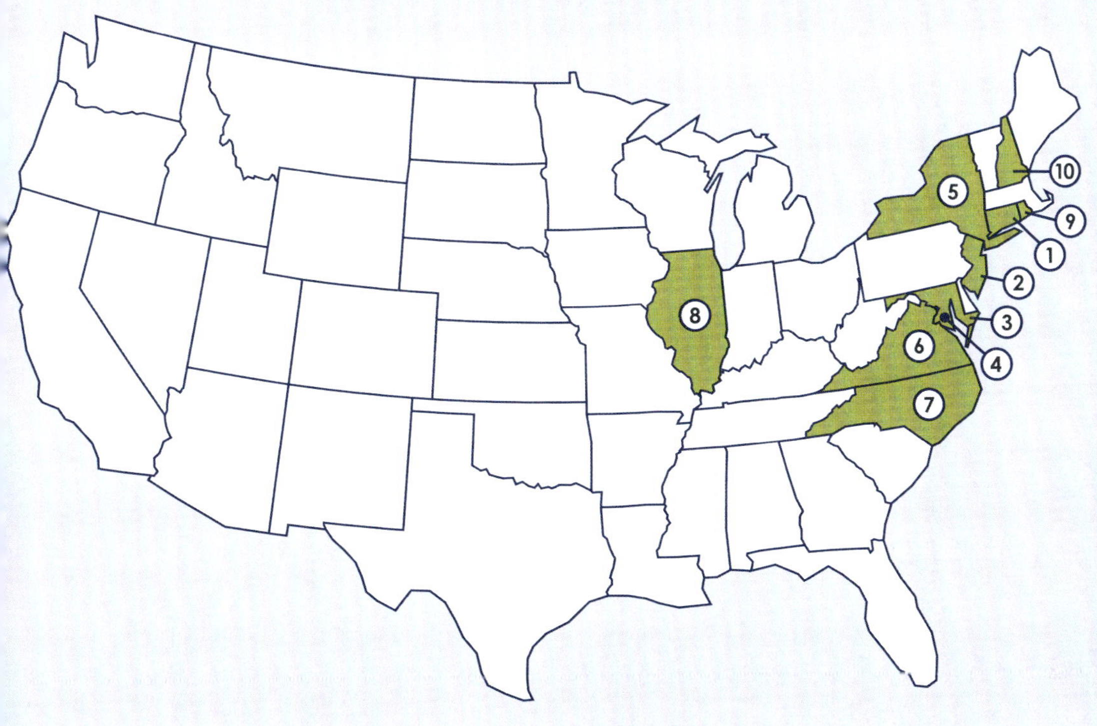

1. Connecticut
2. New Jersey
3. Maryland
4. Washington, DC
5. New York
6. Virginia
7. North Carolina
8. Illinois
9. Rhode Island
10. New Hampshire

Source: John Alpers, "Best States for a Data Analyst," *Zippia*, 2025. www.zippia.com.

These locations offer the most jobs and the best pay for data analysts.

This helps break things up a little. She likes that her position offers this flexibility.

Some data analysts always work in an office. Others always work from home. Remote analysts might work for a company. Or they may do freelance work. This is work companies give to non-employees as needed. Freelance data analysts may work with only one company. Or they could

Staying Connected

Data analysts share information. This is a key part of their job. Even analysts who work from home full time must connect with other people regularly. Analysts need to compare notes with team members. They must provide updates to company clients. They can communicate quickly and easily through virtual meetings.

On platforms such as Fiverr, data analysts can post profiles with samples of their work. Companies looking for help with data analysis can search the site for people who specialize in the type of work they need.

work with several. They may have one or more projects at a time.

Natassha Selvaraj is a data scientist. She helps data analysts begin freelance careers. She suggests they go to online platforms. Upwork and Fiverr are two examples. Many companies post data analytics projects on these sites. Interested freelancers can share their skill sets. Then they can bid on jobs. New analysts can often gain experience by offering to do the work at a low price. This work helps build their résumés. With more experience, they can charge more for their work.

IMPORTANT DATA ANALYTICS TOOLS

A data analyst's main tool is a computer. Analysts use software, too. One such program is Statistical Analysis System (SAS) Studio. It helps analysts manage large amounts of data. It can run code and create reports. It can take data from other software, such as Excel. SAS Studio converts this data so the program can use it. One of SAS's biggest advantages is

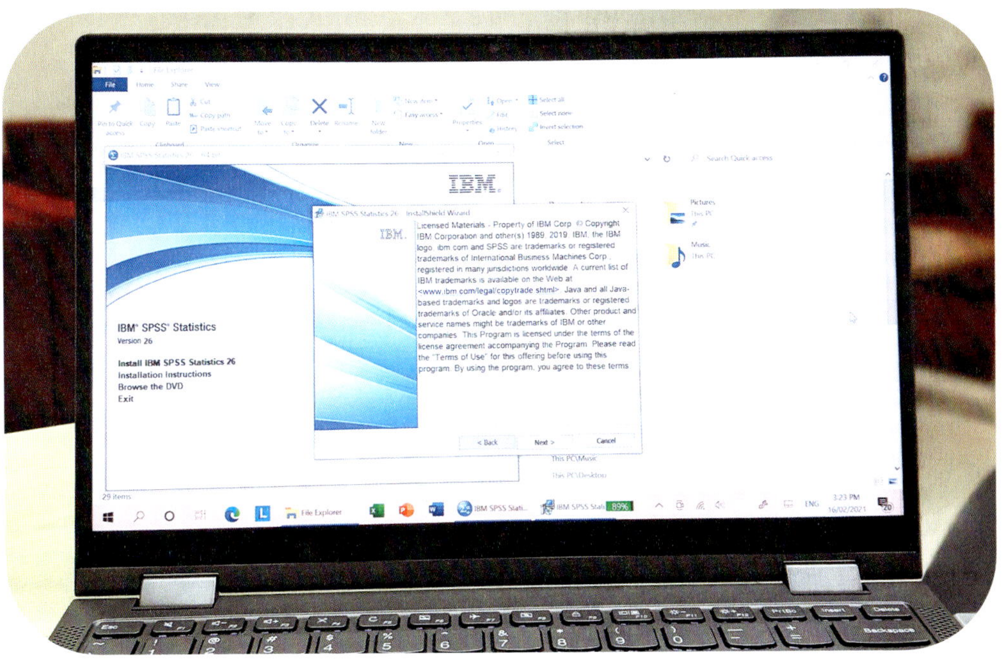

SPSS is a software package offered by IBM. It provides tools that make it easier to analyze, visualize, and predict trends for data.

its security. This makes it ideal for handling **confidential** data.

Data analysts also use Statistical Process for Social Sciences (SPSS). This software is easy to use. It is also reliable. SPSS is often used to study trends. It can be used to test **hypotheses** as well. For example, a government organization might want to study social media to learn how it affects

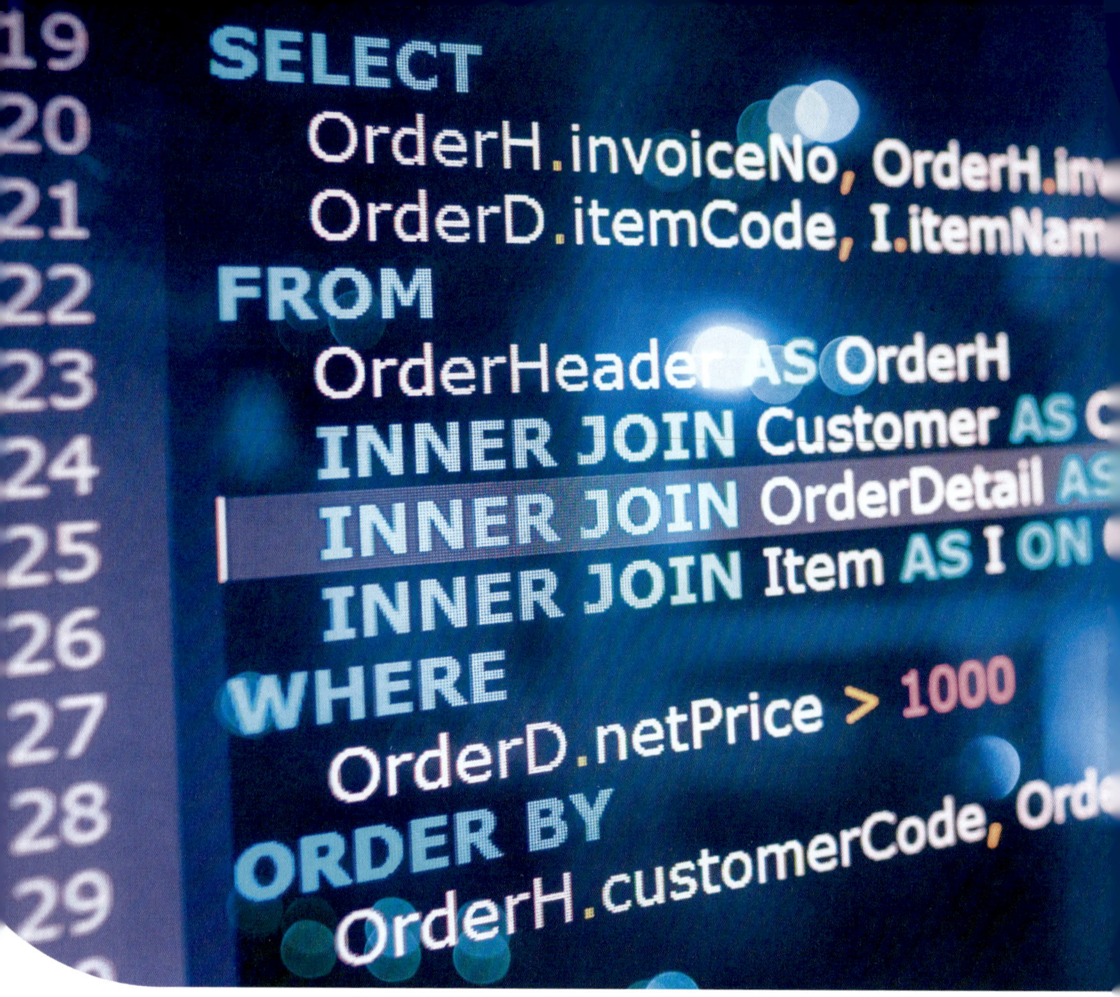

The programming language SQL helps analysts access and use data from databases.

young people. The organization may wonder if those who use social media are lonelier than those who don't. SPSS can help an analyst review data on this topic. It gathers information from surveys teens have taken about social media use.

Knowing programming languages is also important in data analysis. Most data analysts use one or two languages. They may begin by learning SQL. It was created specifically for managing data. It is also easy to learn. SQL helps show sales trends. It also helps analysts see customer behaviors. These may include a customer's common purchases. SQL could also track items that are often returned.

Analysts use Python to find patterns or make predictions. It can also help create graphs or charts of results. R is also popular. It is used by analysts in finance, medicine, and technology.

CHAPTER FOUR

LOOKING AHEAD

Demand for data analysts is on the rise. The Bureau of Labor Statistics (BLS) is a government organization. It gathers data about jobs. The BLS expects this career and field to keep growing. It predicts that demand for data science professionals will increase 36 percent between 2023 and 2033. This means about 20,800 new jobs would become available in the data field each year.

Companies are looking for ways they can use data to be more competitive and efficient. Data analysts play a role in this process.

8.00 9.00 10.00 11.00

AI can improve how data is displayed by creating detailed charts, graphs, and diagrams.

This growth is driven by advances in technology. Artificial intelligence (AI) is the ability of a computer to do work usually done by humans. With the help of AI, companies and governments can gather large amounts of data. Smartphone apps, websites, and social media platforms gather this data. In 2024, more than 2.5 quintillion bytes of data were created daily. This is enough room to store 2.5 trillion books.

DATA ANALYTICS AND AI

Some people think AI will take away data analytics jobs. Experts do not think this will happen. But AI will change the field of analytics. The need for data analysts will likely remain high. What analysts do, though, is likely to change.

In early 2025, AI was already taking over some analytic tasks. This includes

AI gathers data from social media apps such as Instagram, Facebook, and X.

data mining. This is a repetitive job. Giving machines these tasks allows analysts to do more complex jobs.

As AI does more, human data analysts will need to learn to work with AI-driven systems. Analysts will learn to train

Data analysts must review data gathered and presented by AI to ensure it is accurate, consistent, and in line with company values.

AI models. Humans will also interpret data. They are better at this task than AI. Humans will also need to make decisions based on the data. They will need to monitor AI, too. Humans need to make sure AI does not cause harm to people or the planet.

THE WAYS OF THE FUTURE

Up to this point, data analysts focused on managing data. They gathered data. Company leaders then studied the data. Those leaders used it to make business decisions. Future analysts will take a more important role. They will use the data they manage to come up with ideas. The analysts will help solve problems. They will recommend solutions. Some analysts already do this type of work.

In the future, it will become more common for all analysts to provide ideas.

As new tools become available, analysts' jobs will change. Even experienced analysts will need to learn about the latest technology and trends. For example, experts predict that data visualization will change. Advancing technology will allow analysts to use **augmented reality (AR)** and **virtual reality (VR)**. These tools will help analysts show complex data in new ways. Clients may step into imagined worlds. This may help them see possible solutions.

For example, a store owner could use AR or VR to virtually visit a commercial space. They could see two store layout options. The owner could walk through the store

Virtual reality allows data analysts to see and interact with data sets in ways that are not possible when simply viewing them on a screen.

as customers would. They could see how different sales displays stand out. Which layout the owner chooses could have a big effect on future sales.

Analysts must be mindful of using AI ethically. This means respecting user privacy and ensuring the gathered data is free of bias.

Ethics and privacy laws may also play a larger role in data analytics. Analysts must know their limits when mining data. Some states already have data privacy laws.

California, Connecticut, and Virginia have some of the strictest laws. They require smartphone companies to get the user's approval to send data from certain apps to the phone manufacturer.

Companies may have a legal right to certain information. But there may be laws stating how the data can and cannot be used. As technology changes, the laws

Specialized Roles

Data analysts focus on specific duties. This has led to new job titles. Machine learning analysts use data to help computers make simple predictions. And data ethics officers help create rules. These guidelines protect people's rights as companies use their data. Insights strategists develop business plans based on statistics.

about how it can be used may change, too. Analysts will need to stay current on these laws.

WORKING WITH TECHNOLOGY

Dee Radh works in technology marketing. She states:

> *By embracing AI as a tool rather than a threat, data analysts can unlock new levels of productivity and insight, driving smarter business decisions and better outcomes.*[7]

Radh believes AI will help make data analysis even more important than it is now.

Those interested in data analysis do not need degrees. But they must keep learning. Data analysts use math and technology. They also use communication skills.

Data analysts should have a drive to continually learn more about advancements in their field.

They gather and interpret data. This data can fuel the growth of companies. With the help of technology, analysts are in high demand. Many industries need people to perform this important work.

GLOSSARY

augmented reality (AR)

technology that combines computer-generated images with real-world objects and backgrounds

confidential

kept secret

data modeling

the process of creating graphics to help organize data

data visualization

displays such as charts or graphs that summarize the meaning of data

ethics

a set of moral principles that guide human behavior

hypotheses

predictions that can be proven or disproven

mentorship

the sharing of professional knowledge and experience in a one-on-one relationship

virtual reality (VR)

technology that uses computer-generated images and sounds to create an artificial setting or experience

SOURCE NOTES

INTRODUCTION: MAKING THE NUMBERS WORK

1. Avery Smith, "Become a Data Analyst in 90 Days," *YouTube*, uploaded by Avery Smith, October 18, 2023. www.youtube.com

CHAPTER ONE: EXPLORING DATA ANALYTICS

2. Quoted in "Data Analytics in E-Commerce: Definition and Benefits," *Indeed*, July 24, 2025. www.indeed.com.

3. Stephen Baraka, "18 Key Data Analyst Skills to Get You Hired," *Indeed*, June 9, 2025. www.indeed.com.

CHAPTER TWO: TRAINING FOR DATA ANALYTICS

4. Lauren Rosenthal, "Things I Did on My Path to Data Analytics and What I Learned Along the Way," *Medium*, June 15, 2023. https://medium.com.

5. Tamyris Gimenez, "Google Data Analytics Professional Certificate: My Experience," *Medium*, July 1, 2021. https://medium.com.

CHAPTER THREE: WORKING AS A DATA ANALYST

6. Quoted in "Day in the Life of a Data Analyst (In Office)," *YouTube*, uploaded by Agatha, December 15, 2023. www.youtube.com.

CHAPTER FOUR: LOOKING AHEAD

7. Dee Radh, "Will AI Take Data Analyst Jobs?" *Actian*, July 3, 2024. www.actian.com.

FOR FURTHER RESEARCH

BOOKS

Kari Cornell, *Be a Web Designer*. BrightPoint Press, 2025.

Tammy Gagne, *Be a Computer Programmer*. Brightpoint Press, 2026.

George Anthony Kulz, *Jobs in Computer Science*. Abdo Publishing, 2024.

INTERNET SOURCES

"Big Data Science Projects," *Science Buddies*, n.d. www.sciencebuddies.org.

"Data Privacy vs. Data Security," *NOVA*, 2025. https://pbslearningmedia.org.

"Data Scientists," *My Next Move*, n.d. www.mynextmove.org.

WEBSITES

American Statistical Association: Statistics Students

www.amstat.org/education/statistics-students#resources

The Statistics Students page of the American Statistical Association website provides resources that are especially helpful for those starting out in fields that rely on statistical data. It includes links to websites, career resources, and statistics competitions.

Microsoft Excel

www.microsoft.com/en-us/microsoft-365/excel

The Microsoft website provides access to a free version of Microsoft Excel. The site also features videos about how Excel can be used and paired with other software commonly used in the data analysis field.

Raspberry Pi Foundation

www.raspberrypi.org/learn

The Raspberry Pi Foundation is dedicated to teaching young people the skills needed to use technology with confidence. It includes projects and games that help teens learn to code.

INDEX

Amazon, 34, 38
artificial intelligence (AI), 48–51, 56
augmented reality (AR), 52

boot camps, 9, 22, 26–27

certifications, 30–33
classes, 22–27
college degrees, 22, 32, 56
communication skills, 19–21, 40, 56
computer code, 16, 27, 42

data collecting, 12, 14, 16–17
data gathering, 6, 12–14, 16, 19, 48, 51, 57
data interpreting, 14, 18, 51, 57
data managing, 10, 42, 45, 51
data mining, 17, 26, 50, 54
data modeling, 31
data organizing, 18
data visualization, 19, 52
digital marketing, 27

freelance work, 40–42

Google Data Analytics Professional Certificate, 30, 33
Google Sheets, 18, 30

math skills, 18, 56
Microsoft Certified Power BI Data Analyst Associate certification, 31
Microsoft Excel, 9, 18, 24, 25, 26, 42

privacy laws, 54–56
problem solving, 6, 11, 15, 21, 28–29, 34, 37, 51
programming languages, 16, 18, 27, 45
Python, 18, 25, 45
Python Tutorial, The, 28

R, 18, 25, 45

self-taught approaches, 22, 27–30
SQL, 18, 25, 30, 45
Statistical Analysis System (SAS) Studio, 42
Statistical Process for Social Sciences (SPSS), 43–44
statistics, 18, 24, 55

Tableau, 18–19, 25
teamwork, 21, 40
TikTok, 25

Vaporsens, 8–9
virtual reality (VR), 52–53

IMAGE CREDITS

Cover: © AndreyPopov/iStockphoto
5: © Andrey_Popov/Shutterstock Images
7: © Anton Vierietin/Shutterstock Images
8: © Aun Photographer/Shutterstock Images
10: © Undrey/Shutterstock Images
11: © Ground Picture/Shutterstock Images
13: © JLco Julia Amaral/Shutterstock Images
14: © Andrey_Popov/Shutterstock Images
17: © g0d4ather/Shutterstock Images
20: © Summit Art Creations/Shutterstock Images
23: © vitranc/iStockphoto
24: © bangoland/Shutterstock Images
26: © Igor Suka/iStockphoto
28: © Chay_Tee/Shutterstock Images
31: © Smile Studio AP/Shutterstock Images
32: © PeopleImages.com-Yuri A./Shutterstock Images
35: © Tada Images/Shutterstock Images
36: © fizkes/Shutterstock Images
38: © Miljan Zivkovic/Shutterstock Images
39: © The hope/Shutterstock Images
41: © Julio Ricco/Shutterstock Images
43: © Mashka/Shutterstock Images
44: © Tee11/Shutterstock Images
47: © LightField Studios/Shutterstock Images
48: © chayanuphol/Shutterstock Images
49: © Primakov/Shutterstock Images
50: © Andrey_Popov/Shutterstock Images
53: © vectorfusionart/Shutterstock Images
54: © Suri_Studio/Shutterstock Images
57: © Andrey_Popov/Shutterstock Images

ABOUT THE AUTHOR

Tammy Gagne is an author and editor with a passion for educational nonfiction. She has written hundreds of books for both adults and young people. Residing in the beautiful state of Maine, she enjoys life with her husband, their son, and two rescue dogs. When not writing or editing, she is often brainstorming her next project. She hopes her books inspire and educate readers of all ages.